Gustavio Garcia

Brachiaria Brizantha Consortium with Stylosantes Campo Grande

Gustavio Garcia

Brachiaria Brizantha Consortium with Stylosantes Campo Grande

Case study on productivity

ScienciaScripts

Cover image: www.ingimage.com

This book is a translation from the original published under ISBN 978-613-9-65122-1.

Publisher:
Sciencia Scripts
is a trademark of
Dodo Books Indian Ocean Ltd. and OmniScriptum S.R.L publishing group

120 High Road, East Finchley, London, N2 9ED, United Kingdom
Str. Armeneasca 28/1, office 1, Chisinau MD-2012, Republic of Moldova, Europe
Printed at: see last page
ISBN: 978-620-7-85549-0

TABLE OF CONTENTS

SUMMARY

TEODORO,G.G **Productivity of** Brachiaria brizantha **intercropped with** Stylosanthes **Campo Grande** (Stylosanthes capitata x Stylosanthes ***macrocephala*****) with different incoculants.** Araxá - MG 2016

Brazilian livestock farming is based on an extensive production system, using pastures that are almost always exclusively made up of grasses. Estimates suggest that 80% to 90% of the country's pastureland is made up of Brachiaria grass, with B. brizantha occupying more than 50% of the total. However, low meat and milk production rates are observed, and several factors are responsible for this, including low soil fertility and reduced nutrient availability. Among these nutrients, nitrogen is one of the most required by plants, the one that most limits the growth of its aerial part, directly influencing productivity and the one that represents the greatest financial outlay as it accounts for up to 80% of pasture fertilisation costs.

Pastures intercropped with grasses and forage legumes can be a good, low-cost option to alleviate the problem of their degradation, as these legumes are capable of fixing atmospheric N through symbiosis with diazotrophic bacteria, which can improve their quality and productivity.

Therefore, the aim of this study was to assess the productivity of B. brizantha pastures intercropped or not with the legume Estilozantes Campo Grande with and without inoculation with diazotrophic bacteria. The parameters assessed were: green matter productivity ($kg.ha^{-1}$) and dry matter productivity ($kg.ha^{-1}$) of the pasture.

The experiment was conducted in the Fausto de Ávila Experimental Field located on the campus of the Planalto de Araxá University Centre (Uniaraxá) in the municipality of Araxá - MG. The experimental design used was a randomised block design (RBL) with 5 treatments and 4 replications:

T1= Brachiaria brizantha with conventional fertilisation, T2= Brachiria brizantha intercropped with Campo Grande Styloides without inoculation, T3= Brachiaria brizantha inoculated with Azospirillum brasilense, T4= Brachiaria brizantha intercropped with Campo Grande Styloides inoculated with Rhizobium tropici, T5= Brachiaria brizantha intercropped with Campo Grande Styloides inoculated with Bradyrhizobium japonicum.

There was no significant difference at Tukey 5% between any of the treatments for the variables analysed.

Key words: Forage, Pasture, Fertilisation, Biological fixation

PROBIC/UNIARAXÁ and PIBIC/FAPEMIG REGISTRATION FORM

<table>
<tr><th colspan="3">STUDENT DATA</th></tr>
<tr><td colspan="3">Student:Gustavio Garcia Teodoro</td></tr>
<tr><td colspan="3">Programme: Agronomy</td></tr>
<tr><td>Period: 5</td><td colspan="2">Estimated end date: 12/2019</td></tr>
<tr><td colspan="3">Address: Rua Levino Antunes</td></tr>
<tr><td colspan="2">Neighbourhood: Centre</td><td>N°: 770</td></tr>
<tr><td>Date of birth: 14/04/1997</td><td>CPF: 127.439.246-24</td><td>RG: MG-15.043.413</td></tr>
<tr><td>Landline:</td><td colspan="2">Mobile: (34) 9 8895-7462</td></tr>
<tr><td colspan="3">E-mail: gustavioagro@gmail.com</td></tr>
<tr><th colspan="3">SUPERVISOR DETAILS</th></tr>
</table>

Full name of supervisor:Paulo Roberto Fávero de Fravet		
Institute :Uniaraxá		
Work regime: ( ) Hourly (X) Part-time ( ) Full-time		
Title: (x) Master ( ) Doctor		
Address: Rua Terêncio Pereira		
Neighbourhood: Santo Antônio		**N°:2193**
Date of birth: 14/10/1972	**CPF: 16100448866**	**ID: 104791676**
Landline:	**Mobile: (34) 9 9108-6591**	
E-mail: paulofravet@uniaraxa.edu.br		
CO-ORDINATOR'S DETAILS (if any)		
Co-supervisor's full name: Prof. Dr José Carlos da Silva		
Institute: Agricultural and Human Sciences		
Work regime: ( ) Hourly ( ) Part-time (x) Full-time		
Degree: ( ) Master (x) Doctor		
E-mail: josecarlos@uniaraxa.edu.br		
PROJECT DATA		
Project title: Productivity of *Bracmiaria brioantha intercropped with* Estlosantes Campo Grande with different inoculants in the second year of cultivation		
Objectives: To analyse the productivity and quality of pasture of /?rac7/iaria hrízawZ/m intercropped with Stylosantes Campo Grande as a substitute for chemical nitrogen fertilisation.		
Keywords: Biological nitrogen fixation, forage, pasture.		

RESEARCH PROJECT SUBMISSION FORM

1) DATA IDENTIFYING THE RESEARCH PROJECT AND THE PROPOSERS:

***Title of the research project:**	Productivity of *Brachiciria brizantha intercropped with Stylosanthes* Campo Grande (Syy/oswZúes *Capitatn* x *Stylosanthes Mtesocochald)* with different inoculants in a second-year area
***Area of knowledge (CNPq):**	Agronomy
***Sub-area of knowledge (CNPq):**	Phytotechnics

<table>
<tr><td>*Key words:</td><td colspan="2">Biological nitrogen fixation, forage, pasture.</td></tr>
<tr><td rowspan="4">*Nature of the project:</td><td>X</td><td>Scientific Initiation</td></tr>
<tr><td></td><td>Course Conclusion Work</td></tr>
<tr><td></td><td>Extension Project</td></tr>
<tr><td></td><td>Other</td></tr>
<tr><td rowspan="3">*Institutional research line to which the project falls:</td><td></td><td>Environment, Health and Public Policy</td></tr>
<tr><td></td><td>Culture, human development and management</td></tr>
<tr><td>X</td><td>Production, Innovation and Sustainability</td></tr>
<tr><td>*Name of candidate:</td><td colspan="2">Gustavio Garcia Teodoro</td></tr>
<tr><td>*Guidance teacher:</td><td colspan="2">Paulo Roberto Fávero de Fravet</td></tr>
<tr><td rowspan="3">*Supervisor's title:</td><td></td><td>Expertise</td></tr>
<tr><td>X</td><td>Master</td></tr>
<tr><td></td><td>Doctor</td></tr>
<tr><td>*Graduation course to which the project is linked:</td><td colspan="2">Agronomy</td></tr>
</table>

<table>
<tr><td colspan="3">List of researchers participating in the project</td></tr>
<tr><td>Name:</td><td>Maximum qualification:</td><td>Institution to which you are affiliated</td></tr>
<tr><td></td><td></td><td></td></tr>
<tr><td></td><td></td><td></td></tr>
<tr><td></td><td></td><td></td></tr>
<tr><td></td><td></td><td></td></tr>
</table>

* All items marked with (*) on this page are compulsory.

2) DESCRIPTION OF THE RESEARCH PROJECT (SUMMARY OF NO MORE THAN 700 WORDS):

Objective description, supported by specialised literature, of the problem to be studied.

Brazilian livestock farming is based on an extensive production system, using pastures exclusively made up of grasses. Estimates suggest that 80% to 90% of the country's pasture areas are made up of *Brachiaria* grass, and that *B. brizantha* occupies more than 50% of the total made up of these grasses (Boddey etal., 2004). Most of the brachiaria pastures cultivated in different regions of Brazil have shown signs of degradation within a few years of use. The lack of persistence of these grasses can be attributed to

the extractive way in which they are exploited, as well as adaptation problems, inadequate management of the different species, reduced soil fertility, susceptibility to pests and lack of replacement fertilisation (Rodrigues, 1996).
Pastures intercropped with grasses and forage legumes are a good, low-cost option for alleviating the problem of pasture degradation. The use of forage legumes capable of fixing atmospheric N through symbiosis with bacteria of the *Rhizobium* genus can improve pasture quality and provide large quantities of N to the soil-plant-animal system (Giller & Cadisch, 1995; Cantarutti etal., 2002).
The contribution can be made by transferring the N fixed by the legume to the grass, which can increase the support capacity of the pasture and prolong its productive capacity (Cantarutti etal., 2002). In addition, this N fixed by the legume can improve the quality of the diet (Costa, 1995), increasing animal production (Euclides, etal., 1998). Another advantage of intercropping is the lower variation in the nutritional value of legumes compared to forage grasses throughout the year (Jingura etal., 2001).

3) PROJECT OBJECTIVES (200 WORDS MAXIMUM): State the general and specific objectives of the project.

To analyse the productivity (dry matter mass MS) and quality of the pasture, crude protein and NDT of *Brachiaria brizantha* with the use of Campo Grande Stylosantum inoculated with atmospheric nitrogen-fixing bacteria, analysing the possibility of reducing the use of chemical nitrogen fertilisers, avoiding contamination of the soil and especially the water table with the inappropriate use of this fertiliser and reducing dependence on external inputs, making it possible to reduce costs with pasture fertilisation. Dry mass productivity and forage quality will be analysed for each treatment.

4) JUSTIFICATION FOR THE PROJECT (MAXIMUM 200 WORDS):

Describe the relevance and specific importance for the development of the project.

Brazil's pastures should be seen as a crop, just as corn, soya and others are, because from pastures we harvest meat and milk for the animals that will consume them. Chemical nitrogen fertilisation accounts for 80% of pasture fertilisation costs, as well as causing soil and water contamination. The consortium between grasses and legumes could replace part of this chemical fertilisation, so this project could help clarify doubts about the best way to fix nitrogen and meet the needs of meat and milk producers in the Araxá region who use *Brachiria brizantha* as pasture for their livestock.

5) METHODOLOGY (MAXIMUM 500 WORDS):

Describe the methodology that will be used to develop the project (how the objectives will be achieved).

The experiment will be conducted in the Fausto de Ávila experimental field of the University Centre of Planalto de Araxá - UNIARAXÁ, located on Avenida Ananias Teixeira Aguiar, whose soil consists of a medium-textured Yellow Red Latosol. Sowing will take place in February 2016, using *Brachiaria brizantha* seeds as indicated by the manufacturer. Before sowing, the peanut seeds will be inoculated with nitrogen-fixing bacteria.

The statistical design will be randomised blocks*, with* 5 treatments and 4 replications:

Treatment 1: *Brachiaria brizantha* with conventional fertiliser.

Treatment 2: *Brachiaria brizantha* inoculated with Azospirillum.

Treatment 3: *Brachiaria brizantha* intercropped with Stylosantes Campo Grande.

Treatment 4: *Brachiaria brizantha intercropped* with Stylosantes Campo Grande inoculated with Rhizobhium SP.

Treatment 5: *Brachiaria brizantha intercropped* with Stylosantes Campo Grande inoculated with Bradyrhizobium japonicum.

Each treatment will be harvested using a rake whenever the *Brachiaria brizantha* reaches a height of 40 cm, which is the height indicated for animals to enter this pasture. After each harvest, the dry matter of the harvested material will be determined using the microwave method and then weighed on a precision scale to determine productivity over the course of the year. The green mass will be sent to a laboratory specialised in determining pasture quality, FERTILAB, located in the municipality of São Gotardo-MG, Rod. MG 235 Km 89,700 - Zona Rural, CEP - 38800000, Caixa Postal 163, where bromatological analysis, crude protein, NDT and sterile extract will be carried out.

6) EXPECTED RESULTS (MAXIMUM 200 WORDS):

Describe the results expected from the development of the project (production of knowledge and/or development of products and/or services, with publications in journals, presentations at scientific events, filing and/or registration of patents).

The experiment is expected to clarify whether it is possible to reduce chemical nitrogen fertilisation with the intercropping used in the experiment. And which would be more viable with the productivity and quality of the *Brachiaria brizantha.*

7) PROJECT EXECUTION SCHEDULE (MAXIMUM 200 WORDS):

Describe the proposed work plan, with the steps to be taken.

	Ja n	Fe V	Ma r	Ab r	Ma i	Ju n	Ju 1	Ag 0	If t	Or t	In V	Fro m z	Ja n
Review of Hist.			X	X	X	X	X	X	X	X	X	X	X
NDT analysis				X				X					
Crude protein analysis				X				X					
Measurement Analysis				X		X		X		X		X	

8) BUDGET FOR THE IMPLEMENTATION OF THE PROJECT (MAXIMUM 200 WORDS):

Describe the estimated costs for carrying out the project.

Laboratory Analysis: R$ 350.00
String: R$ 60,00

9) REFERENCES (MAXIMUM 500 WORDS):

Cite the references used to prepare the project (check that the citations are relevant, sufficient and up-to-date).

ANDRADE, C.M.S.; GARCIA, R.; COUTO, L. et al. Factors limiting the growth of tanzania grass in an agroforestry system with eucalyptus, in the Cerrados region of Minas Gerais. Revista Brasileira de Zootecnia, v.30, n.4, p.1178-1185, 2001 (supplement)

BARCELOS, A.O.; VILELA, L.; MARTHA JÚNIOR, G.B. Using protein banks as an alternative for supplementing dairy cows. In: SYMPOSIUM ON TECHNOLOGICAL, PROCESSUAL AND PUBLIC POLICY ALTERNATIVES FOR SUSTAINABLE MILK PRODUCTION, 5, 2003, Juiz de Fora. Proceedings. Juiz de Fora: Embrapa Gado de Leite, 2003. p.93-119.

BODDEY, R.M.; MACEDO, R.; TARRÉ, R.; FERREIRA, E.; OLIVEIRA, O.C.; RESENDE, C. de P.; CANTARUTTI, R.B.; PEREIRA, J.M.; ALVES, B.J.R.; URQUIAGA, S. Nitrogen cycling in Brachiaria pastures: the key to understanding the process of pasture decline. Agriculture, Ecosystems and Environment, v.103, p.389-403, 2004.

CANTARUTTI, R.B.; TARRÉ, R.M.; MACEDO, R.; CADISCH, G.; RESENDE, C.P.; PEREIRA, J.M.; BRAGA, J.M.; GOMEDE, J.A.; FERREIRA, E.; ALVES, B.J.R.; URQUIAGA, S.; BODDEY, R.M. The effect of grazing intensity and the presence of a forage legume on nitrogen dynamics in Brachiaria pastures in the Atlantic forest region of the South of Bahia, Brazil. Nutrient Cycling in Agroecosystem, v.64, p.257-271, 2002.

COSTA, N.L. Nitrogen fertilisation and intercropping of elephant grass (Pennisetum purpureum cv. Cameroon) with tropical forage legumes. Pesquisa Agropecuária Brasileira, v.30, p.401-408, 1995.

EUCLIDES, V.P.B.; CARDOSO, E.G.; MACEDO, M.C.M.; OLIVEIRA, M.P. Voluntary intake of Brachiaria decumbens cv. Basilisk and Brachiaria brizantha cv. Maradu under grazing. Revista Brasileira de Zootecnia, v.29, p.2200-2208, 2000. (Supplement)

EUCLIDES, V.P.B.; MACEDO, M.C.M.; OLIVEIRA, M.P. Cattle production on Brachiaria spp. pastures intercropped with Calopogonium mucunoides in the Cerrados. Revista Brasileira de Zootecnia, v.27, p.238-245, 1998.

GILLER, K.; CADISCH, G. Future benefits from biological nitrogen fixation: an ecological approach to agriculture. Plant and Soil, v.174, p.255-277, 1995.

JINGURA, R.M.; SIBANDA, S.; HAMUDIKUWANDA, H. Yield and nutritive value of tropical forage legumes grown in semi-arid parts of Zimbabwe. Tropical Grassland, v.35, p.168-174, 2001.

LASCANO, C.E.; EUCLIDES, V.P.B. Nutritional quality and animal production of Brachiaria pastures. In: MILES, J.W.; MAASS, B.L.;

VALLE, C.B. (Ed.). Brachiaria: biology, agronomy and improvement. Cali: CIAT;
Campo Grande: Embrapa-CNPGC, 1996. p.106-123
PEREIRA, J.M.; NASCIMENTO JUNIOR, D.; SANTANA, J.R.; CANTARUTTI, R.B.; LEÃO, M.I. Crude protein content and "in vitro" dry matter digestibility of available forage and of the diet selected by cattle grazing B. humidicola (Rendle) Scweick, in monoculture or intercropped with legumes, subjected to different stocking rates. Revista da Sociedade Brasileira de Zootecnia, v.21, p. 104-117, 1992.
QUEIROZ, A.C.; SILVA, D.J. Food analysis: chemical and biological methods. Viçosa: Federal University of Viçosa, 2002. 235p.
RENNÓ, L.N. Intake, total and partial digestibility, microbial production, rumen parameters and urea and creatinine excretion in steers fed diets containing four levels of urea or two levels of protein. Viçosa, MG: Federal University of Viçosa, 2003. 252p. Thesis (Doctorate in Animal Science) - Federal University of Viçosa, 2003
RODRIGUES, L.R.A.; RODRIGUES, T.J.D. Establishment of Cynodon grass in areas of Brachiaria spp. In: WORKSHOP ON THE FORAGE POTENTIAL OF THE CYNODON GENUS, 1996, Juiz de Fora. Proceedings. Juiz de Fora: Embrapa Gado de Leite, 1996. p.8-21.
VALADARES FILHO, S.C.; VALADARES, R.F.D. Recent advances in protein in dairy cow nutrition. In: INTERNATIONAL SYMPOSIUM
DE BOVINOCULTURA DE LEITE, SINLEITE, 2., 2001, Lavras. Proceedings... Lavras: Federal University of Lavras, 2001. p.228-243.
VALLE, C.B.; EUCLIDES, V.P.B.; MACEDO, M.C.M. Characteristics of forage plants of the genus Brachiaria. In: SIMPÓSIO SOBRE MANEJO DE PASTAGEM, 17., 2001, Piracicaba. Proceedings. Piracicaba: Fealq, 2001. p.133-176.

10) ETHICAL ASPECTS (IF THE RESEARCH INVOLVES STUDIES WITH HUMAN BEINGS OR EXPERIMENTAL ANIMALS):

10.1) LETTER OF INFORMATION TO THE RESEARCH SUBJECT (MAXIMUM 300 WORDS):

Reproduce the letter, stating the objectives, the justification, the procedures, as well as the risks, discomforts and benefits of the procedures and the conditions for the research subject's participation.

10.2) INFORMED CONSENT FORM (MAXIMUM 200 WORDS):
Reproduce the Informed Consent Form to be used in the study.

10.3) CONDITIONS OF MAINTENANCE AND EUTHANASIA, IN THE CASE OF STUDIES USING EXPERIMENTAL ANIMALS (MAXIMUM 300 WORDS):
Describe the maintenance and euthanasia conditions to be used.

11) AWARENESS OF THOSE INVOLVED IN THE RESEARCH PROJECT:

Signature of the researchers involved in carrying out the research project.

Date:	Full name:	Signature:
06/03/2017	Paulo Roberto Fávero de Fravet	
06/03/2017	Gustavio Garcia Teodoro	

12) OPINION OF THE COURSE COLLEGIATE TO WHICH THE PROJECT IS LINKED

Opinion issued by the Course Board after assessment of the research proposal.

	Approved by the Course Board
	Rejected by the Course Committee

Comments from the Course Board (maximum 200 words):

Signature and stamp of the Course Coordinator

13) OPINION OF THE RESEARCH, EXTENSION AND POSTGRADUATE AFFAIRS BOARD

Opinion issued by the Research, Extension and Postgraduate Affairs Board, after the research proposal has been assessed by the Course Board.

	Approved by the Course Board
	Rejected by the Course Committee

Comments from the Course Board (maximum 200 words):

Signature and stamp of the President of the Research, Extension and Postgraduate Affairs Board

CHAPTER 1

INTRODUCTION

Brazil is currently at the forefront of livestock farming, as it is a country whose climate favours the production of fodder in different locations and at different times of the year. It stands out as the largest commercial producer of cattle in the world, where the activity is one of the main factors responsible for the country's economic growth (BENETT & BUZETTI, 2007). The current Brazilian herd is approximately 209 million head and pasture has been used as the main source of food for these animals (it is estimated that only 3% of the herd is finished in an intensive system) (ABIEC, 2009).

According to the Brazilian Institute of Geography and Statistics (IBGE - 2010), there are a total of 174 million hectares of pastures in Brazil, and Minas Gerais has an area of more than 25 million hectares of pastures (native and planted), which represents 43 per cent of its entire territory.

The biggest problems facing livestock farming in Portugal are related to the degradation of our pastures, which results in low quality, low productivity and consequently low carrying capacity.

The main factors causing this degradation include inadequate management practices, inappropriate fertilisation and biotic problems such as attacks by insect pests.

Choosing the right species of forage is one way of minimising this degradation. Grasses of the Brachiaria genus are the most common in our country, accounting for around 80% of pastures.

As previously mentioned, the inadequate fertilisation of pastures is one of the main causes of their degradation, with nitrogen (N) accounting for around 80% of the costs of this practice.

The consortium between grasses and legumes could replace part of this chemical fertilisation, so this project could help clarify doubts about the best way to fix nitrogen and meet the needs of meat and milk producers in the Araxá region who use Brachiria brizantha as pasture for their livestock.

CHAPTER 2

OBJECTIVES

2.1 GENERAL OBJECTIVE

To evaluate the influence of intercropping with the legume Estilozantes Campo Grande and inoculation with different diazotrophic bacteria on the productivity of B. brizantha pasture.

2.2 SPECIFIC OBJECTIVES

Analyse the following pasture parameters:

- Green matter yield (kg.ha);$^{-1}$
- Dry matter yield (kg.ha);$^{-1}$

CHAPTER 3

THEORETICAL FRAMEWORK

3.1 Brazilian livestock farming

Brazil is currently the world's second largest producer and largest exporter of beef. Practically all of Brazil's beef production is based on pastures, the most economical and practical way of producing and providing food for cattle. Pastures therefore play a fundamental role in ensuring low production costs (DIAS FILHO 2014).

An important characteristic of Brazilian livestock farming is that most of its cattle are raised on pasture (FERRAZ; FELÍCIO, 2010), which is the most economical and practical way to produce and provide food for cattle. As a result of this vocation of Brazilian livestock farming, mainly due to the country's climatic characteristics and territorial extension, Brazil has one of the lowest meat production costs in the world (CARVALHO et al., 2009; DEBLITZ, 2012; FERRAZ; FELÍCIO, 2010).

3.2 The genus Brachiaria

In Brazilian agriculture, the Brachiaria genus has been considered an instrument for including the cerrado in the production process. This grass, which is one of the best known and most widely used forage plants today, was introduced to the cerrado as a means of cultivating low-fertility soils which, due to their acidity, presented serious nutritional restrictions for crops. Today, the various species of this genus are internationally recognised for their contribution to the advancement of Brazilian livestock farming, breaking adaptation barriers and setting production records, allowing for significant gains in stocking rates, performance and animal productivity (COSTA, 2006).

Its unquestionable importance for Brazilian agriculture comes from its good nutritional value for livestock and its adaptability to different ecosystems.

This includes characteristics such as tolerance to drought, cold and leafhoppers, depending on the cultivar. In order to create a good pasture, in addition to good quality seeds, it is essential to choose the right *Brachiaria* species and cultivar, depending on the purpose of use, soil and climate characteristics, the investments you intend to make in the area, among others (LACERDA, 2007).

3.2.1Brachiaria *brizantha*

Brachiaria brizantha originated in tropical Africa and South Africa. It was first studied by the National Beef Cattle Research Centre (CNPGC-EMBRAPA) and later by the National Centre for Agricultural Research in the Cerrados (CNPAC-EMBRAPA).

Its size is highly variable, as is its pubescence and yield. It thrives in most soils, including acidic ones (WENZL et al., 2002), but requires more than 500 mm of rain per year. It is almost erect, roots very little at the nodes and adapts to more or less humid regions, from sea level to over 3,000 metres above sea level. It is moderately drought tolerant, grows well in non-moist soils, is cold tolerant, resistant to leafhopper attack, has good fodder value and high green mass production and low seed production.

3. 3Forage legumes

Legumes offer various uses for those who want to cultivate the soil with respect for the environment. They utilise nitrogen from the atmosphere, benefiting the plants that are intercropped with them and improving soil quality. It is recommended that they be intercropped with forage grasses, interspersed with perennial plants (such as coffee and citrus) or used for green manure. The introduction of legumes to pasture can increase animal production by increasing the quality and quantity of forage on offer. Thus, the results are visible not only in the animal's diet, but also in the indirect effects related to biological nitrogen fixation and its transfer to the pasture ecosystem. (Embrapa, 1996)

3.4 *Stylosanthes* Campo Grande *(Stylosanthes capitata x Stylosanthes macrocephala)*

The Campo Grande *Stylosanthes* legume is a cross between two legume species: *Stylosanthes macrocephala* and *Stylosanthes capitata.* It grows semi-erect and can reach heights of up to 1.5 metres. It has good soil cover and high forage quality, but the pasture takes a while to establish itself. The high Cultural Value (70) of its seed allows for high productivity and high natural reseeding capacity. Campo Grande is recommended for cattle grazing (CARDOSO; VALENTIM, 1982; VALENTIM, 1983).

Campo Grande *Stylosanthes* Seed *(macrocephala x capitata)* is a hybrid of two legume species, *Stylosanthes macrocephala* and *Stylosanthes capitata.* The *capitata* usually grows erect and can reach heights of up to 1.5 metres, producing rounded leaves and flowers that vary in colour from yellow to beige.

Macrocephala, on the other hand, grows a little more horizontally, producing pointed leaves and yellow flowers. Its height is similar to that of the *capitata,* but it matures and flowers a little faster than the *capitata.* Stylosantes Campo Grande has good ground cover and high forage quality, but it takes a long time to establish the pasture. It is recognised as an excellent nitrogen fixer in the soil and has a high protein value. Given these characteristics, this legume adapts easily to sandy, low-fertility soils. It also has a high seed yield and good digestibility. Recommended for cattle grazing (ANDRADE; VALENTIM, 2007).

3.5 BIOLOGICAL NITROGEN FIXATION

Nitrogen fixation is carried out by diazotrophic bacteria. When planting, the seeds are placed in contact with the bacteria in a process called inoculation. Once this is done, the bacterium waits for germination and penetrates the roots of the legume, where it develops nodules and lodges in them, with the two organisms living in harmony. In this form of life in partnership, called symbiosis, the bacterium finds the ideal environment for its existence, feeding on substances provided by the host plant, without causing damage, and fixes nitrogen, favouring the plant in its development and production (EMBRAPA 2014).

N is easily found in the soil, in water, in living beings and especially in atmospheric air. Around 78 per cent of the air in the atmosphere is made up of nitrogen. However, even with this large amount of nitrogen available in the environment, many or almost all crops are deficient in this nutrient, due to the fact that only 0.000 per cent of it is readily available to plants. The other large part needs to undergo chemical reactions in order to be made available (HAVLIN, 2005).

The efficiency of Biological Nitrogen Fixation (BNF) is affected by a series of factors such as temperature, soil acidity, the relationship between bacteria and host, soil moisture, soil compaction, etc. (MNASRI et al., 2007).

3.6 Diazotrophic bacteria

Diazotrophs are organisms capable of fixing atmospheric nitrogen (N2), i.e. they are able to transform nitrogen from an unavailable form (N2) into an available form (ammonium, i.e. NH4+).

The group of organisms that receives this name is made up of microorganisms, more specifically bacteria that can live in the wild or in association with other organisms. In the latter case, the association between bacteria of the genus Rhizobium and leguminous plants is well known. While the bacteria provide nitrogen for the plant, the latter provides organic material (glucose) produced through photosynthesis. So, in simple terms, the plant helps the bacteria and the bacteria helps the plant in an association called mutualism.

As mentioned, there are free-living diazotrophs, i.e. they fix atmospheric nitrogen on their own for their own use without needing to be associated with other organisms, unlike the case just presented. These bacteria carry out what is known as non-symbiotic nitrogen fixation. In this case, for example, bacteria of the genus Azotobacter are very common in the soil. You should also be aware that there are specific bacteria for each plant species, such as Bradyrhizobium japonicum for soya and Rhizobium tropici for beans (Salemi L.F., 2010).

3.6.1 Azospirillum brasilense

Bacteria of the genus Azospirillum became known worldwide in the mid-70s. Several studies have shown the ability of these bacteria to fix atmospheric N when associated with poaceae (formerly the grass family) (TARRAND et al., 1978).

These bacteria produce phytohormones that stimulate the development of the roots of various plants (TIEN et al., 1979). Barassi et al., (2008), showed positive changes in poaceae resulting from inoculation with Azospirillum, such as improved phytosynthetic parameters in the leaves (chlorophyll content), improved cell wall elasticity, increased biomass production and greater plant height. Correa et al. (2008) also observed greater absorption of water and minerals by plants inoculated with this bacterium, making them even more resistant to adverse weather conditions.

It is essential that the selection of strains of Azospirillum spp. with high BNF potential in association with forage grasses be intensified, and new studies be carried out by conducting experiments both in the greenhouse and in the field, so that strains can be selected that are capable of supplying part of the nitrogen demand required by pastures. In this way, and with proper management of these areas, forage species can be grown sustainably, without generating environmental degradation (BALDANI et al., 2000; GUIMARÃES et al., 2010).

3.6.2 Rhizobium tropici

Rhizobium is a genus of gram-negative bacteria native to the soil of the American continent and they fix atmospheric nitrogen, making them essential in the nitrogen cycle. Bacteria of the genus Rhizobium are endosymbionts and live in the cells of legume roots. These bacteria colonise cells within the root nodules of the plant's roots, where they convert atmospheric nitrogen into ammonia (this process is called biological nitrogen fixation (BNF) or ammonification), after which the bacteria supply the plant with organic nitrogen compounds such as glutamine or ureides. The plants in turn provide the bacteria with organic compounds produced through photosynthesis. Inoculation of the bean plant with this bacterium represents a reduction of up to 20% of all nitrogen fertilisation (CANTARELLA, 2007).

Studies aimed at increasing bean productivity through the symbiotic fixation of atmospheric nitrogen have shown the possibility of obtaining bean yields of up to 1,600 kg ha^{-1} (Dobereiner & Duque, 1980).

3.6.3Bradyrhizobium Japonicum

These bacteria native to Asia are capable of fixing atmospheric nitrogen in symbiosis with legumes. In Brazil, the species Bradyhizobium japonicum has been widely used in soya bean crops (Glicine max) (EMBRAPA, 2007.

They are Gram-negative and the optimum temperature for their growth is between 25 and 30°C (HOLT et al., 1994; SOMASEGARAN and HOBEN, 1994).

Inoculation of bacteria from the Bradyrhizobium genus in soya beans provides a considerable reduction in the cost of producing this crop by eliminating the need for nitrogen fertilisation, while also avoiding negative effects on the environment due to the leaching process of nitrogenous forms, polluting rivers and lakes, and by denitrification, affecting the ozone layer (EMBRAPA, 2001).

3. 7Dry Matter (DM)

Determining dry matter is the starting point for analysing forage crops. It is of great importance since the preservation of the food may depend on the moisture content present in the material and, furthermore, when comparing the nutritional value of two or more foods, we have to take into account their respective dry matter contents. This is the weight of the material analysed free of water. It is important to know the percentage of dry matter contained in the sample, as this is the basis for calculating the diet, since feed consumption by animals is expressed in kg of dry matter per animal (RODRIGUES, 2010).

The microwave method (EMBRAPA, 2007) was used to determine dry matter using the following procedures:

1. Weigh a plastic tray with approximate dimensions of 20x20 cm (weight A);
2. Add 80 to 100 g of forage and weigh (weight B);

3. Dry in a domestic microwave oven, using the following heating schedule: 3 min at 20 per cent of maximum power, 10 min at 100 per cent of maximum power and 5 min at 50 per cent of maximum power;

4. Remove the tray from the oven, weigh, homogenise the material and heat again for 1 min at maximum power;

5. Remove the tray from the oven again and weigh the dry sample (weight c); repeat operations 4 and 5 until the weight of the sample is constant;

6 Calculate the dry matter (DM) using the equation: DM(%)= (C-A) X 100/(B-A)

CHAPTER 4

METHODOLOGY

The experiment was conducted in the Fausto de Ávila experimental field located on the campus of the Planalto de Araxá University Centre - UNIARAXÁ in the city of Araxá-MG. The soil consists of a medium-textured Yellow Red Latosol and its chemical and physical properties are shown in the table below.

pH WATERA	mg/dm³		Cmolc/dm³							%		dag/kg	
	P	K	AI	Ca	Mg	H+Al	SB	t	T	V	M	M.O-	C
6,1	8,2	90, 0	0, 0	4, 3	1,6	2,4	6,1	6,1	8,5	71, 9	0,0	3,6	2,1

Source: Atelpe Laboratory

Sowing was carried out on 1 March 2016, using Brachiaria brizantha and Campo Grande Stilisantes seeds as indicated by the manufacturer. Moments before sowing, the seeds of treatments 2, 4 and 5 were inoculated with the specific diazotrophic bacterium. The seeds were planted using the no-till system and the planting fertiliser was 200 kg.ha^{-1} of Super Simple Phosphate in the sowing line, which was the same for all the treatments. Both seeds were sown in a row in 5 cm deep furrows with 14 seeds per linear metre. Treatment 1 received a single top dressing of 200 kg.ha^{-1} of urea, applied by throwing 30 days after the Brachiaria seedlings emerged.

The experimental design used was a randomised block design consisting of 5 treatments and 4 replications and each plot was 3m x 2m, totalling $6m^2$.

Treatment 1: Brachiaria brizantha with conventional fertiliser.

Treatment 2: Brachiaria brizantha inoculated with Azospirillum.

Treatment 3: Brachiaria brizantha intercropped with Stylosantes Campo Grande.

Treatment 4: Brachiaria brizantha intercropped with Stylosantes Campo Grande inoculated with Rhizobhium SP.

Treatment 5: Brachiaria brizantha intercropped with Stylosantes Campo Grande inoculated with Bradyrhizobium japonicum.

On 17/10/2017, the treatments were harvested using a rake when the Brachiaria brizantha reached a height of 40 cm, leaving a residual of 20 cm to the ground. After harvesting, the material was weighed on a precision scale to determine the green matter and then taken to the laboratory to determine the dry matter content using the micro-wave method. All the results were statistically analysed using SASM-AGRO software at a significance level of 5% using the Tukey test, and the data was tabulated in spreadsheets and consolidated into tables.

CHAPTER 5

RESULTS AND DISCUSSIONS

Tables 1 and 2 show that there was no significant difference between the treatments in terms of green matter yield (Kg.ha^{-1}) and dry matter yield (Kg.ha^{-1}), respectively.

Table 1. Green Matter Production (kg.ha^{-1}) as a result of the intercropping of Brachiaria Brizantha with Campo Grande Styloxanthus with different inoculants. Araxá - MG (2016).

TREATMENTS	**Green Matter (MV)** (Kg.ha)$^{-1}$
T1	**16.768 a**
T2	**16.816 a**
T3	**15.564 a**
T4	**12.296 a**
T5	**17.738 a**

SOURCE: Prepared by the author
CV: 20.87%
*Means followed by the same letters in the column do not differ according to Tukey's test at 5% probability.

Using the microwave method, the dry matter production shown in Table 2 was evaluated.

Table 2. Dry Matter Production (kg.ha^{-1}) as a function of the intercropping of Brachiaria Brizantha with Campo Grande Stylosants with different inoculants. Araxá - MG (2016).

TREATMENTS	**Dry Matter (DM)** (Kg.ha-1)
T1	**6.254,5 a**
T2	**6.339,6 a**
T3	**5.797,6 a**
T4	**4.911,6 a**

T5	5.955,5 a

SOURCE: Prepared by the author
CV: 9.17%
*Means followed by the same letters in the column do not differ according to Tukey's test at 5% probability.

According to Salemi, (2010), there are specific bacteria for each plant species, so it is likely that the bacteria Rhizobium tropici and Bradyrhizobium japonicum did not influence biological nitrogen fixation in the Campo Grande Styloxanthus, which probably symbiotised with some other diazotrophic bacteria in the soil, since the result of treatment T3 was statistically the same as treatments T4 and T5.

The result of the T2 treatment agrees with the work carried out by Hungria et al. (2010) in Londrina and Ponta Grossa analysing strains of Azospirillum brasilense in wheat, where they found a 31% increase in the productivity of this grass.

CHAPTER 6

CONCLUSION

The results show that there was no significant difference in the variables Green Matter and Dry Matter of the pasture between the treatments evaluated at Tukey 5%, under the conditions of this study.

This result shows us that it is possible to partially or even totally replace (in this case, up to 200 $kg.ha^{-1}$ of urea) chemical nitrogen fertilisation through a consortium between Brachiaria brizantha and Campo Grande Stylosants or by inoculating Brachiaria brizantha with Azospirilum brasiliense. It also indicates that there is a need to study which diazotrophic bacteria have the best results for symbiosis with Campo Grande Styloxanthus. This favours the productivity and quality of the country's pastures, the livestock farmers' economy and everyone's environment.

CHAPTER 7

REFERENCES

ANDRADE, C.M.S.; GARCIA, R.; COUTO, L. et al. Factors limiting the growth of tanzania grass in an agroforestry system with eucalyptus, in the Cerrados region of Minas Gerais. Revista Brasileira de Zootecnia, v.30, n.4, p.1178-1185, 2001 (supplement)

BARCELOS, A.O.; VILELA, L.; MARTHA JÚNIOR, G.B. Using protein banks as an alternative for supplementing dairy cows. In: SYMPOSIUM ON TECHNOLOGICAL, PROCESSUAL AND PUBLIC POLICY ALTERNATIVES FOR SUSTAINABLE MILK PRODUCTION, 5, 2003, Juiz de Fora. Proceedings. Juiz de Fora: Embrapa Gado de Leite, 2003. p.93-119.

BODDEY, R.M.; MACEDO, R.; TARRÉ, R.; FERREIRA, E.; OLIVEIRA, O.C.; RESENDE, C. de P.; CANTARUTTI, R.B.; PEREIRA, J.M.; ALVES, B.J.R.; URQUIAGA, S. Nitrogen cycling in Brachiaria pastures: the key to understanding the process of pasture decline. Agriculture, Ecosystems and Environment, v.103, p.389-403, 2004.

CANTARUTTI, R.B.; TARRÉ, R.M.; MACEDO, R.; CADISCH, G.; RESENDE, C.P.; PEREIRA, J.M.; BRAGA, J.M.; GOMEDE, J.A.; FERREIRA, E.; ALVES, B.J.R.; URQUIAGA, S.; BODDEY, R.M. The effect of grazing intensity and the presence of a forage legume on nitrogen dynamics in Brachiaria pastures in the Atlantic forest region of the South of Bahia, Brazil. Nutrient Cycling in Agroecosystem, v.64, p.257-271,

Clima Temperado, 2010. 177p. - (Embrapa Clima Temperado. Documentos, 306). ISSN 1516-8840.

COSTA, N.L. Nitrogen fertilisation and intercropping of elephant grass (Pennisetum purpureum cv. Cameroon) with tropical forage legumes. Pesquisa Agropecuária Brasileira, v.30, p.401-408, 1995.

EUCLIDES, V.P.B.; MACEDO, M.C.M.; OLIVEIRA, M.P. Cattle production on Brachiaria spp. pastures intercropped with Calopogonium mucunoides in the Cerrados. Revista Brasileira de Zootecnia, v.27, p.238-245, 1998.

EUCLIDES, V.P.B.; CARDOSO, E.G.; MACEDO, M.C.M.; OLIVEIRA, M.P. Voluntary intake of Brachiaria decumbens cv. Basilisk and Brachiaria brizantha cv. Maradu under grazing. Revista Brasileira de Zootecnia, v.29, p.2200-2208, 2000. (Supplement)

GILLER, K.; CADISCH, G. Future benefits from biological nitrogen fixation: an ecological approach to agriculture. Plant and Soil, v.174, p.255-277, 1995.

JINGURA, R.M.; SIBANDA, S.; HAMUDIKUWANDA, H. Yield and

nutritive value of tropical forage legumes grown in semi-arid parts of Zimbabwe. Tropical Grassland, v.35, p.168-174, 2001.

LASCANO, C.E.; EUCLIDES, V.P.B. Nutritional quality and animal production of Brachiaria pastures. In: MILES, J.W.; MAASS, B.L.; VALLE, C.B. (Ed.). Brachiaria: biology, agronomy and improvement. Cali: CIAT; Campo Grande: Embrapa-CNPGC, 1996. p.106-123

RODRIGUES, L.R.A.; RODRIGUES, T.J.D. Establishment of Cynodon grass in areas of Brachiaria spp. In: WORKSHOP ON THE FORAGE POTENTIAL OF THE CYNODON GENUS, 1996, Juiz de Fora. Proceedings. Juiz de Fora: Embrapa Gado de Leite, 1996. p.8-21.

RODRIGUES, L.R.A.; RODRIGUES, T.J.D. Establishment of Cynodon grass in areas of Brachiaria spp. In: WORKSHOP ON THE FORAGE POTENTIAL OF THE CYNODON GENUS, 1996, Juiz de Fora. Proceedings. Juiz de Fora: Embrapa Gado de Leite, 1996. p.8-21.

RODRIGUES, R, C.; Methods of bromatological food analysis: physical, chemical and bromatological methods / Ruben Cassel Rodrigues. - Pelotas: Embrapa

RODRIGUEZ, N.M.; CAMPOS, W.E.; LACHICA, M.L. et al. A calorimetry system for metabolism trials. Arquivo Brasileiro de Medicina

Veterinaria e Zootecnia, v.59, n. 2, p. 495-500, 2007.

OUYANG, D.S.; MACKENZIE, A.F. & FAN, M.X. Ammonia volatilisation from urea amended with triple superphosphate and potassium chloride. Soil Sci. Soc. Am.J., 62:1443-1447, 1998.

OUYANG, D. et al. Phytotoxicity of banded urea amended with triple superphosphate and potassium chloride. Agronomy Journal, Madison, v.90, p.734-739, 1998.

PIRES, W. **Manual de Pastagem: fomação, manejo e recuperação.** Editora Aprenda Fácil, Viçosa, 2006, 304p.

PRIMAVESI, A. C. et al. Nitrogen fertilisation in Coastcross grass: effects on nutrient extraction and apparent nitrogen recovery. **Revista Brasileira de Zootecnia**, Viçosa, v. 33, n. 1, p. 68-78, 2004.

RODRIGUES, L.R.A.; RODRIGUES, T.J.D. Establishment of Cynodon grass in areas of Brachiaria spp. In: WORKSHOP ON THE FORAGE POTENTIAL OF THE CYNODON GENUS, 1996, Juiz de Fora. Proceedings. Juiz de Fora: Embrapa Gado de Leite, 1996. p.8-21.

RODRIGUES, R, C.; Methods of bromatological food analysis: physical, chemical and bromatological methods / Ruben Cassel Rodrigues. - Pelotas: Embrapa Clima Temperado, 2010. 177p. - (Embrapa Clima Temperado. Documentos, 306). ISSN 1516-8840.

RODRIGUEZ, N.M.; CAMPOS, W.E.; LACHICA, M.L. et al. A calorimetry system for metabolism trials. Arquivo Brasileiro de Medicina Veterinaria e Zootecnia, v.59, n. 2, p. 495-500, 2007.

PRIMAVESI, A. C. et al. Nitrogen fertilisation in Coastcross grass: effects on nutrient extraction and apparent nitrogen recovery. **Revista Brasileira de Zootecnia**, Viçosa, v. 33, n. 1, p. 68-78, 2004.

PEREIRA, J.M.; NASCIMENTO JUNIOR, D.; SANTANA, J.R.; CANTARUTTI, R.B.; LEÃO, M.I. Crude protein content and "in vitro" dry matter digestibility of available forage and of the diet selected by cattle grazing B. humidicola (Rendle) Scweick, in monoculture or intercropped with legumes, subjected to different stocking rates. Revista da Sociedade Brasileira de Zootecnia, v.21, p.104-117, 1992.

QUEIROZ, A.C.; SILVA, D.J. Food analysis: chemical and biological methods. Viçosa: Federal University of Viçosa, 2002. 235p.

RODRIGUES, L.R.A.; RODRIGUES, T.J.D. Establishment of Cynodon grass in areas of Brachiaria spp. In: WORKSHOP ON THE FORAGE POTENTIAL OF THE CYNODON GENUS, 1996, Juiz de Fora. Proceedings. Juiz de Fora: Embrapa Gado de Leite, 1996. p.8-21.

RODRIGUES, R, C.; Methods of bromatological food analysis: physical, chemical and bromatological methods / Ruben Cassel Rodrigues. - Pelotas: Embrapa Clima Temperado, 2010. 177p. - (Embrapa Clima Temperado. Documentos, 306). ISSN 1516-8840.

RODRIGUEZ, N.M.; CAMPOS, W.E.; LACHICA, M.L. et al. A calorimetry system for metabolism trials. Arquivo Brasileiro de Medicina Veterinaria e Zootecnia, v.59, n. 2, p. 495-500, 2007.

SANTOS, M.E.R.; FONSECA, D.M.; EUCLIDES, V.P.B. et al. Cattle production on deferred brachiaria grass pastures. Revista Brasileira de Zootecnia, v.38, n.4, p.635- 642. 2009 a

VALLE, C.B.; EUCLIDES, V.P.B.; MACEDO, M.C.M. Characteristics of forage plants of the genus Brachiaria. In: SIMPÓSIO SOBRE MANEJO DE PASTAGEM, 17., 2001, Piracicaba. Proceedings. Piracicaba: Fealq, 2001. p.133-176.

VITOR, C. M.T.; FONSECA, D. M.; COSER, A. C. et al. Dry matter production and nutritive value of elephant grass pasture under irrigation and nitrogen fertilisation. **Revista Brasileira de Zootecnia**, Coronel Pacheco, v.38, n.3, p.435-442, 2009.

VILELA, H. **Pasture: Selection of forage plants, establishment and fertilisation.** Aprenda Fácil Editora, Viçosa-MG, 2005, 283p.

WERNER, J.C. **Fertilisation of pastures**. Nova Odessa: Instituto de Zootecnia, 1986. 49p. (Technical Bulletin, 18).

ZIMMER, A.H.; MACEDO, M.C.M.; BARCELLOS, A.O.; KICHEL, A.N. Establishment and recovery of Brachiaria pastures. In: Peixoto, A.M.; Moura, J.C.; Faria, V.P. (eds.). SYMPOSIUM ON PASTURE MANAGEMENT, 11, Piracicaba, 1994. Proceedings... Piracicaba: FEALQ, 1994. 325p.

ZIMMER, A.H.; EUCLIDES FILHO, K. Pastures and Brazilian beef cattle farming. In: Gomide, J.A. (ed.). INTERNATIONAL SYMPOSIUM ON ANIMAL PRODUCTION IN PASTURE, 1, 1997, Viçosa. Proceedings...

ZIMMER, A. R.; EUCLIDES, V. P. B.; MACEDO, M. C. M. Management of forage plants of the Brachiaria genus. In: SYMPOSIUM ON PASTURE MANAGEMENT. 1988. Piracicaba, Proceedings of the symposium on pasture management. Piracicaba: FEALQ, 1988. p. 141-183.

MIX
Papier aus verantwortungsvollen Quellen
Paper from responsible sources
FSC® C105338

Printed by Books on Demand GmbH, Norderstedt / Germany